INTERNATIONAL CENTRE FOR MECHANICAL SCIENCES

COURSES AND LECTURES - No. 43

GERHARD HEINRICH
TECHNICAL UNIVERSITY OF VIENNA

GAS-LUBRICATED BEARINGS OF GYROSCOPES

COURSE HELD AT THE DEPARTMENT
FOR GENERAL MECHANICS
SEPTEMBER - OCTOBER 1970

UDINE 1970

SPRINGER-VERLAG WIEN GMBH

This work is subject to copyright.

All rights are reserved,

whether the whole or part of the material is concerned

specifically those of translation, reprinting, re-use of illustrations,

broadcasting, reproduction by photocopying machine

or similar means, and storage in data banks.

© 1972 by Springer-Verlag Wien

Originally published by Springer-Verlag Wien-New York in 1972

ISBN 978-3-211-81147-4   ISBN 978-3-7091-2718-6 (eBook)
DOI 10.1007/978-3-7091-2718-6

Foreword

This monograph is a somewhat concise presentation of a series of six lectures I delivered in October 1970 at CISM in Udine.

Air-bearings are used to-day in many fields of technology. They have stood the test with machine-tools and many different appliances. In my lectures I dealts with air-bearings for gyroscopes. During World War Two the use of air-bearings was a pioneer work. It was then that I had the chance to do research work in the development of air-bearings for gyroscopes. I have remained in touch with that field, producing papers on its problems from time to time.

The subject matter is comparatively easy from the theoretical point of view. The attractive side of it lies in its direct application to technological problems. My lectures were delivered with a view to show how to attain relevant statements that can be made use of in the field of construction.

In the introduction static and dynamic air-bearings for gyroscopes are dealt with. The second part discusses the laminar and turbulent field of a viscose, compressible medium through a gap and the flow through an orifice in the range of sub-sonic and sound velocity. The third part treats the double operating thrust-bearing and the static journal bearing. The fourth and last part is dedicated to the self-supporting air-bearing and to self-excited instabilities that may prop up.

It is my pleasant duty to record my sincere thanks to the authorities of CISM for their invitation to deliver these lectures. In particular, I want to thank Professor Luigi Sobrero whose advice and ready help have always been of greatest value. Not less have I to thank my dear friends, Professor W. Olszak and Professor H. Parkus, without whose initiative I could scarcely have come into touch with CISM. Last but not least, I want to give my thanks to Dr. H. Troger, a member of my institute, for his valuable assistance at the presentation of my lectures.

October 1970					G. Heinrich

1. Introduction.

The support of gyroscopes by gas- or air lubricated bearings has a number of advantages. We have, however, to distinguish the support by gimbal bearings from the one by spin-axis. (Figure 1).

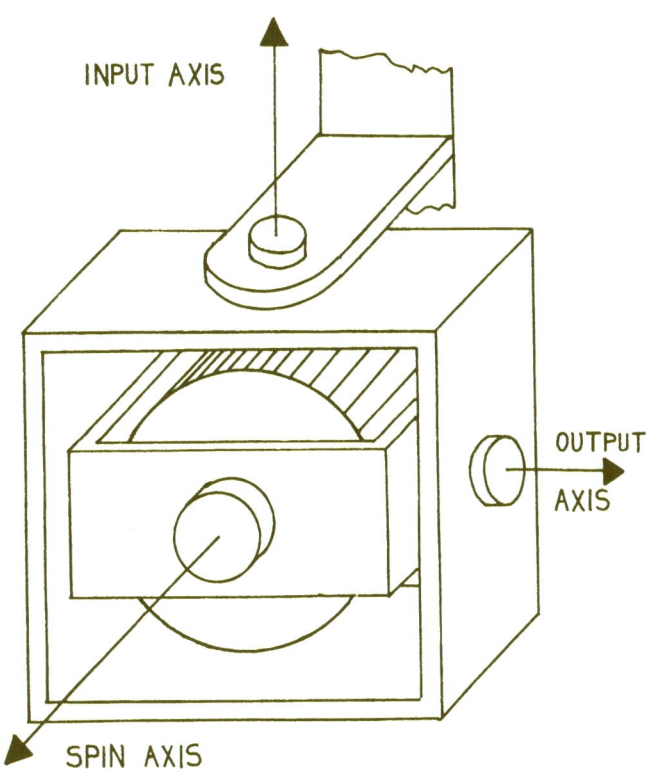

Figure 1.

The friction torques in the gimbal bearings effect the motion of the gyroscope as erratic torques, and we try, therefore, to keep them as small as possible. The viscosity of gases is lower than that of liquid lubricants by several orders of magnitude. As the friction torques are proportional both to the viscosity of the lubricant and to the sliding velocity of the bearing, the friction is practically eliminated with gas-lubricated bearings of low sliding velocities as it is the case with gimbal bearings. On top of that we have the favourable condition that, unlike with liquid lubricants, the viscosity of gases increases with rising temperatures, so that they are suited as lubricants even with higher temperatures.

For the support of the spin axis the gas lubricated bearings also possess great advantages over the ball bearings. Self-acting dynamical gas lubricated bearings are used in this case. They will be discussed later on more thoroughly. The following qualities of ball bearings of the spin-axis may contribute to gyroscope error.

1. Dynamical reactions of the balls and the retainer can impose erratic torques to the gyroscope rotor.

2. Uncontrolled migration of the lubricating oil can contribute to a shift of the mass centre relative to the output gimbal-axis.

3. Lubrication deterioration after prolonged running can aggravate erratic dynamical ball-retainer reac-

tions and can also increase bearing friction. Excessive bearing friction can cause the rotor to drop out of synchronism.

    4. Gyroscope ball bearings provide for both radial and axial stiffness simultaneously. If the axial and radial compliances of the spin axis bearing should be unequal, an error torque, known as the compliance torque, would be produced.

    The gas-bearing gyroscope, because the mass of the gas is negligible as compared to the inertial effects of the gyroscope rotor, and because the lubricating properties of gases do not change with running, is free of the first three types of errors described above. The compliance torque problem, however, also exists with the gas-bearing gyroscope; in fact, it is more complicated because the displacement of the self-acting gas-lubricated journal bearing usually has a component perpendicular to the applied load. This matter will be discussed in more detail later.

2. Parallel Flow in the Gap.

    We shall start now to work out some fundaments for the so-called static gas-bearings, as they are used for gymbal bearings. We can distinguish thrust bearings from journal bearings, according to the supporting force lying in the direction of the axis or perpendicularly to it. A characteristical factor of the statical gas-bearings consists in the gases and the air being supplied by orifices, which are fed from a container. From the orifices the gas enters into the gap of the bearing and flow through it. In consequence of the excess pressure buoyancy is effected in a resting bearing. The movement with gymbal bearings is negligible.

An important foundation for the theory of the gas flow in statical gas bearings is the investigation into a rectilinear parallel flow through a gap (Fig. 2).

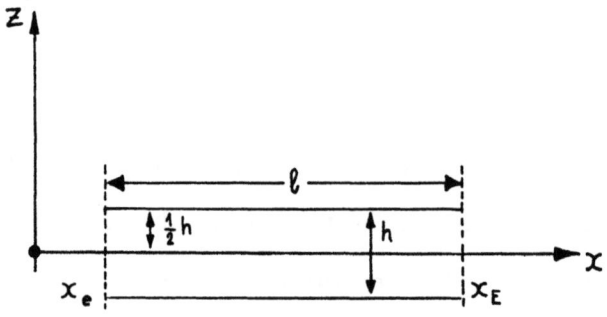

Figure 2.

The gap shall have the width b in the direction y the length ℓ in the direction x and the thickness h in the direction z.

We suppose :

$$-\frac{1}{2}b \leq y \leq \frac{1}{2}b ,$$

$$-\frac{1}{2}h \leq z \leq \frac{1}{2}h$$

Parallel Flow in the Gap

and
$$l > b \gg h.$$

In consequence of this inequality, the flow through the gap can, apart from boundary disturbances, be looked at as a plane flow in the x-z-plain. Because of the inequality the initial process at the entrance $(x = x_e)$ can be neglected. Because of the metallic boundary walls and the good heat-convection connected with it, the flow, in conformity with the experiment may be regarded as isothermal. First air will be used as gas.

The validity of the equation of state for ideal gases

$$\frac{p}{\varrho} = RT. \tag{1}$$

With the individual gas-constant R can be assumed for the gas For the isothermal change of state holds $T = T_0 = \text{const}$. For the laminar flow of a viscose liquid, the equation of Stokes-Navier holds. It is in vectorial representation:

$$\varrho \frac{D\vec{u}}{dt} + \nabla p = \eta \left[\Delta \vec{u} + \frac{1}{3} \nabla (\nabla \cdot \vec{u}) \right]. \tag{2}$$

In it $\vec{u}$ is the vector of velocity, η the absolute viscosity and ∇ the well-known vector operator, $\frac{D}{dt}$ means the substantial differential quotient. Finally the equation of continuity holds:

$$\nabla \cdot (\varrho \vec{u}) + \frac{\partial \varrho}{\partial t} = 0 \tag{3}$$

For the plane problem in question there holds, in case of a stationary flow, the initial process being neglected:

$$u_x = u_x(x,z) \ ; \quad u_y = u_z = 0 \ ;$$

$$\frac{Du_x}{dt} = u_x \frac{\partial u_x}{\partial x} \ ; \quad \frac{Du_z}{dt} = 0 \ ; \quad \nabla \cdot \vec{u} = \frac{\partial u_x}{\partial x} \ .$$

Eq. (2) therefore reads in components:

$$\varrho u_x \frac{\partial u_x}{\partial x} + \frac{\partial p}{\partial x} = \eta \left(\frac{\partial^2 u_x}{\partial z^2} + \frac{4}{3} \frac{\partial^2 u_x}{\partial x^2} \right),$$

$$\frac{\partial p}{\partial z} = \eta \cdot \frac{1}{3} \frac{\partial}{\partial z} \left(\frac{\partial u}{\partial x} \right) .$$

Finally the equation of continuity reads: $\varrho \frac{\partial u_x}{\partial x} + u_x \frac{\partial \varrho}{\partial x} = 0$.

If we assume a thin gap $(l \gg h)$ the pressure becomes independent from z and we get:

$$\frac{\partial}{\partial z} \left(\frac{\partial u_x}{\partial x} \right) = 0 \ .$$

As there exists a great dependence on z because of the condition of adherance along the walls we can neglect $\frac{\partial u_x}{\partial x}$ in the equations of Stokes-Navier: the same holds for: $\frac{\partial^2 u_x}{\partial x^2}$.

We now write: $u_x = u$. We therefore get:

(4) $$\frac{\partial p}{\partial x} = \eta \frac{\partial^2 u}{\partial z^2} \ .$$

As because of the thin gap p only depends upon x we can write: $\frac{\partial p}{\partial x} = \frac{dp}{dx}$. Equation (4) can be integrated with respect to z, x being constant, and we get the boundary condi

Parallel Flow in the Gap

tions $z = \pm \dfrac{h}{2} \ldots u = 0$: taken into account :

$$u = \frac{1}{2\eta} \frac{dp}{dx} (z^2 - \frac{h^2}{4}) \, .$$

If we take the mean value of u along the cross-section :

$$\bar{u} = \frac{1}{h} \int_{-\frac{h}{2}}^{+\frac{h}{2}} u \, dz \, ,$$

we get :

$$\frac{dp}{dx} = -12\eta \, \frac{\bar{u}}{h^2} \, . \tag{5}$$

In this equation $\bar{u}$ will depend upon x.

From the equation of continuity (3) there results, the flow being stationary : $\varrho \cdot \bar{u} = \text{const}$; ϱ is constant along the cross-section together with p because of the equation of state (1), the flow being isothermal. If we take the mean value there results :

$$\varrho(x) \cdot \bar{u}(x) = \text{const} \, . \tag{6}$$

Consequently there results from (1) :

$$p(x) \cdot \bar{u}(x) = \text{const} \, . \tag{7}$$

(5) and (7) give :

$$-\frac{h^2}{12\eta} \cdot p \cdot \frac{dp}{dx} = \text{const} \, .$$

The flow being isothermal $\eta = $ const too, therefore

$$p \cdot \frac{dp}{dx} = \text{const}$$

holds true the thickness of the gap being constant, unlike the incompressible flow, where:

$$\frac{dp}{dx} = \text{const} .$$

The gas supply to the gap is performed through orifices as shown in Figure 3.

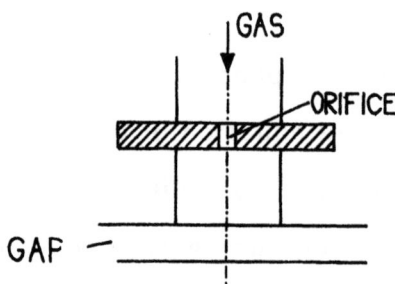

Figure 3.

In this case the gas, or the air respectively, passes through a narrow orifice which gives into a stabilizing chamber, from which the gas enters into the gap. Because of the small length of the orifice the flow through it can be assumed as frictionless. In this case the change of state however, is not isothermal but adiabatic, i.e. there is no thermical exchange with the surroundings.

At first we shall generally denote the quantity

Parallel Flow in the Gap

of heat fed to the gas per unity of mass, the temperature rising for dT with $c_n \cdot dT$ where c_n is a specific heat, the quantity of which depends upon the kind of the change of state. According to the first law of thermodynamics we get

$$c_n \cdot dT = c_v \cdot dT + p \, d\left(\frac{1}{\varrho}\right).$$

For the adiabatic change of state we have: $c_n = 0$; using the equation:

$$c_p - c_v = R \qquad (8)$$

we get, taking into account the equation of state (1):

$$p \cdot d\left(\frac{1}{\varrho}\right) + \frac{dp}{\varrho} = (c_p - c_v) \cdot dT.$$

When eliminating dT we get finally:

$$d\left(\frac{1}{\varrho}\right) \cdot p \, \frac{c_p - c_n}{c_v - c_n} = -\frac{dp}{\varrho},$$

and from it by integration: $\dfrac{p}{\varrho^n} = \text{const.}$, with

$$n = \frac{c_p - c_n}{c_v - c_n}. \qquad (9)$$

For the adiabatic change of state we get:

$$n = \frac{c_p}{c_v} = \varkappa, \qquad (10)$$

and therefore: $\dfrac{p}{\varrho^\varkappa} = \text{const}.$ \qquad (11)

As long as n remains open we speak of a polytropic change

of state. For adiabatic changes of state we have :

$$c_v dT + p d\left(\frac{1}{\varrho}\right) = 0 .$$

According to the equation of state (1) in connection with (8) we get :

$$p d\left(\frac{1}{\varrho}\right) = (c_p - c_v) dT - \frac{dp}{\varrho}$$

and therefore :

(12) $$c_p dT = \frac{dp}{\varrho} .$$

For a frictionless flow ($\eta = 0$) there follows from (2) Euler's equation :

(13) $$\frac{D\mathring{u}}{dt} = -\frac{1}{\varrho}\nabla p = 0 .$$

In a stationary one-dimensional flow there is :

$$\frac{D\bar{u}}{dt} = \bar{u}\frac{d\bar{u}}{dx}$$

where $\bar{u}$ is the velocity of flow averaged across the cross-section, and x the coordinate in the direction of the axis. Therefore there follows from (13) :

(14) $$\bar{u} d\bar{u} + \frac{dp}{\varrho} = 0$$

and according to (12) : $\bar{u}^2 = 2c_p(T_0 - T)$; T_0 is the temperature for $u = 0$ Using (8) and (10) we get :

$$\bar{u} = \sqrt{\frac{2\varkappa}{\varkappa - 1} R (T_0 - T)} . \qquad (15)$$

From (11) we get by means of the equation of state (1) :

$$\frac{T}{p^{\frac{\varkappa-1}{\varkappa}}} = \text{const.}, \text{ respectively} \qquad \frac{T}{T_0} = \left(\frac{p}{p_0}\right)^{\frac{\varkappa-1}{\varkappa}} . \qquad (16)$$

Finally we have :

$$\frac{p}{\varrho^\varkappa} = \text{const.}, \text{ respectively} \qquad \frac{p}{p_0} = \left(\frac{\varrho}{\varrho_0}\right)^\varkappa . \qquad (17)$$

For the density of the mass flow there follows the relation :

$$\varrho\bar{u} = \sqrt{\varrho_0 \frac{2\varkappa}{\varkappa - 1} RT_0 \left[\left(\frac{p}{p_0}\right)^{\frac{2}{\varkappa}} - \left(\frac{p}{p_0}\right)^{\frac{\varkappa-1}{\varkappa}}\right]} . \qquad (18)$$

If the final pressure in the orifice is p_E the cross-section of the orifice A_E and α the orifice coefficient, then, the mass flow per second $\frac{dm}{dt}$ through the orifice is given by :

$$\frac{dm}{dt} = \alpha A_E \frac{p_0}{\sqrt{RT_0}} \sqrt{\frac{2\varkappa}{\varkappa - 1} \left[\left(\frac{p_E}{p_0}\right)^{\frac{2}{\varkappa}} - \left(\frac{p_E}{p_0}\right)^{\frac{\varkappa-1}{\varkappa}}\right]} . \qquad (19)$$

The relation between the flow- and sound-velocity c is of fundamental importance. It is given by :

$$c = \sqrt{\frac{dp}{d\varrho}} \qquad (20)$$

is therefore dependent upon the change of state.
The differential of the density of the mass-flow

$$d(\varrho\bar{u}) = \bar{u}\,d\varrho + \varrho\,d\bar{u}$$

gives, together with Euler's equation (14):

(21) $$d(\varrho\bar{u}) = \left(\bar{u}^2 - \frac{dp}{d\varrho}\right)\frac{d\varrho}{\bar{u}} .$$

One sees from it that $\varrho\bar{u}$ assumes an extreme value (maximum) when reaching the sound velocity c according to eq. (13). As, according to the equation of continuity $\varrho\bar{u}A$ must be constant, (A the respective cross-section of the orifice), there follows: $A = \dfrac{C}{\varrho\bar{u}}$.

A must, therefore, be a minimum when reaching the sound-velocity. If we do not use a Laval orifice, which widens after having reached the sound-velocity, this velocity can be reached but not transgressed. With gas-bearings Laval orifices are not used. From (17) and (20) we get the well-known value of the adiabatic sound-velocity:

(22) $$C_{ad} = \sqrt{\frac{\varkappa p}{\varrho}} = \sqrt{\varkappa RT} .$$

If T_{cr} is the temperature at which c_{ad} is reached, then we get from (15), (16) and (22):

(23) $$T_{cr} = \frac{2T_c}{\varkappa + 1} .$$

Parallel Flow in the Gap

From (16) we get the respective "critical pressure":

$$p_{cr} = p_0 \left(\frac{2}{\varkappa+1}\right)^{\frac{\varkappa}{\varkappa+1}} . \qquad (24)$$

The opposite pressure in the exit chamber p_g must satisfy the inequality:

$$p_g \leq p_E \qquad (25)$$

where p_E is the final pressure in the orifice. The following cases are possible: If $p_g > p_E$ then, $p_E > p_{cr}$ because of (25); the sound-velocity can not be reached. As the exit of the gas takes place with subsonic velocity there is: $p_E = p_g$. If however, $p_g < p_{cr}$ then, p_E can be equal to p_{cr}, the gas reaches the sound-velocity and there is $p_E > p_g$.

In the gap itself the sound-velocity can be reached but not transgressed. The change of state in the gap is isothermal therefore the sound-velocity c_{is} in the gap is also isothermal. According to eq. (20) it becomes:

$$c_{is} = \sqrt{RT_0} . \qquad (26)$$

From (14) we get together with the isotherm: $\frac{p}{\varrho} = RT_0$ the relation:

$$\bar{u}\,d\bar{u} = -RT_0 \frac{dp}{p}$$

and from it by integration :

(27)
$$\bar{u}^2 = RT_0 \ln\left(\frac{p_0}{p}\right)^2 .$$

When the isothermal sound-velocity is reached the respective critical pressure p_{cr} becomes $p_{cr} = \frac{p_0}{\sqrt{e}}$, independent from the nature of the gas. For air $\varkappa = 1,4$ and therefore $p_{cr} = 0,53\, p_0$ in case of a diabatic change of state, and $p_{cr} = 0,61\, p_0$ in case of an isothermal change of state.

With the flow in the parallel gap we have neglected the inertial term $\varrho u_x \frac{\partial u_x}{\partial x}$ in the equation of Stokes-Navier. Now we can discuss to which quantity of flow-velocity this neglection is allowed. From the equation of continuity :

$$\varrho \frac{\partial u_x}{\partial x} + u_x \frac{\partial \varrho}{\partial x} = 0 .$$

follows for the inertial term : $-u_x^2 \frac{\partial p}{\partial x}$. In the case of isothermal change of state we get : $-\frac{u_x^2}{RT_0} \frac{\partial p}{\partial x}$. Substituting this term into the equation of Stokes-Navier we get :

$$\left(-\frac{u_x^2}{RT_0} + 1\right) \frac{\partial p}{\partial x} = \eta\left(\frac{\partial^2 u_x}{\partial z^2} + \frac{4}{3}\frac{\partial^2 u_x}{\partial x^2}\right)$$

$\frac{u_x^2}{RT_0}$ is small compared with 1 always when the flow-velocity is small compared with the isothermal sound-velocity.

Up-to-now we have assumed that the one-dimensional flow in the gap is a laminar one. In the case of a turbulent flow we can only speak of a stationary flow if we av-

Parallel Flow in the Gap

erage over the time. Besides the average over the time of the flow-velocity u is almost independent from z. In case of turbulence the equation of motion can be written as follows:

$$\bar{\rho}\bar{u}\frac{d\bar{u}}{dx} + \frac{dp}{dx} = -\frac{\lambda \bar{\rho}\bar{u}^2}{4h} \tag{28}$$

λ is the resistance coefficient, the friction term is generally $\frac{\lambda \bar{\rho}\bar{u}^2}{2d_H}$ where $d_H = \frac{4A}{U}$ is the hydraulic diameter, $A = b \cdot h$ is the cross-section flow, and $U = 2(b+h)$ is the moistened circumference, whence $d_H = 2h$ for $b \gg h$.

The equation of continuity reads here: $\frac{d}{dx}(\rho\bar{u}) = 0$ in case of isothermal change of state we have: $\rho = \frac{p}{RT_0}$ and therefore:

$$\frac{d}{dx}(p\bar{u}) = 0. \tag{29}$$

With $\rho = \frac{p}{RT_0}$ in connection with (28) and (29) we get:

$$\frac{dp}{dx} - \frac{\bar{u}^2}{RT_0} \cdot \frac{dp}{dx} = -\frac{\lambda}{4}\frac{\bar{u}}{RT_0}\cdot\frac{p}{h},$$

or after multiplication with $2p \cdot dx$:

$$2p \cdot dp - 2\frac{dp}{p}\frac{(p\bar{u})^2}{RT_0} = -\frac{\lambda}{2}\frac{(p\bar{u})^2}{RT_0}\frac{dx}{h}. \tag{30}$$

From the equation of state (29) follows:

$$p u = p_e p_e = p_E p_E \tag{31}$$

where p_e and u_e respectively p_E and u_E are the values of p and u in the points x_e and x_E.

Integrating (30) we get :

(32) $$p_e^2 - p^2 - \frac{(p_e \cdot \bar{u}_e)^2}{RT_0} \ln\left(\frac{p_e^2}{p^2}\right) = \frac{\lambda}{2} \frac{(p_e \cdot \bar{u}_e)^2}{RT_0} \frac{x - x_e}{h}$$

and

(33) $$p_e^2 - p_E^2 - \frac{(p_e \bar{u}_e)^2}{RT_0} \ln \frac{p_e^2}{p_E^2} = \frac{\lambda \ell}{2h} \frac{(p_e \bar{u}_e)^2}{RT_0}$$

with $x_E - x_e = \ell$.

If there is the gas plenum pressure p_0 before the gap and the opposite pressure p_g behind the gap, we have:

$$p_0 > p_e > p_E \gtreqqless p_g .$$

Whether a laminar or a turbulent flow develops is decided by Reynolds' number, as is well known. It is given by : $R_e = \frac{\varrho \bar{u} d_H}{\eta}$ or if $d_H = 2h$ then :

(34) $$R_e = 2h \frac{\varrho \bar{u}}{\eta} .$$

If Reynolds' number transgresses a critical value $R_{e\,cr}$ the flow becomes turbulent.

In order to get a relation that possesses the same form both for the laminar and for the turbulent flow we compare eq. (5) for the laminar flow with eq. (28) for the turbulent one. If we substitute for η, according to eq. (34) : $\eta = \frac{2h\varrho\bar{u}}{R_e}$ in eq. (5), we get :

$$\frac{dp}{dx} = -\frac{96 \varrho \bar{u}^2}{R_e 4h} \qquad \text{for the laminar flow .}$$

The comparison with (28) yields, if we cancel

Parallel Flow in the Gap

the inertial term:

$$\lambda_{lam} = \frac{96}{R_e} \quad . \tag{35}$$

For the flow in the laminar field through a circular tube we get, as is well known: $\lambda_{lam} = \frac{64}{R_e}$. Here the critical Reynolds' number is to be found at about $R_{e\,cr} = 2320$. Experiments have shown that, with the flow through the gap we have to take into account a critical Reynolds' number of

$$R_{e\,cr} = 2690 \quad . \tag{36}$$

The respective value of λ_{cr} follows from (35) as $\lambda_{cr} = 0{,}036$. The measurements show furtheron that in the field of $2960 \leq R_e \leq 8000$ the value of λ remains almost constant, with

$$\lambda = \lambda_{cr} = 0{,}036 \quad . \tag{37}$$

For the laminar flow the equations (32) and (33) read therefore, if we consider (34), (35) and the equation of continuity (6), as well as the equation of state: $\varrho_e = \frac{p_e}{RT_0}$:

$$p_e^2 - p^2 - \frac{(p_e \bar{u}_e)^2}{RT_0} \ln\left(\frac{p_e^2}{p^2}\right) = \frac{24\eta}{h^2}(p_e \bar{u}_e)(x - x_e) \tag{38}$$

$$p_e^2 - p_E^2 - \frac{(p_e \bar{u}_e)^2}{RT_0} \ln\left(\frac{p_e^2}{p_E^2}\right) = \frac{24\eta l}{h^2}(p_e \bar{u}_e) \quad . \tag{39}$$

So far the essential connections for the rectilinear parallel flow in the gap are given.

With gyroscopes one does not use high feeding pressures for the gas supply to the gymbal bearings in general, so that the flow in the gap can be assumed as a laminar one. If $\dfrac{p}{p_e}$ does not diverge much from one the flow through the orifices allows to be substituted by an approximate formula. Instead of (18) we get an approximation :

$$(40) \qquad \varrho \bar{u} = \sqrt{2\varrho_0(p_0 - p)}$$

and, instead of (19) :

$$(41) \qquad \frac{dm}{dt} = \alpha A_E \sqrt{2\varrho_0(p_0 - p_E)} .$$

3. Application to Statical Gas Bearings.

At first we consider thrust-bearings supported in two directions. (figure 4).

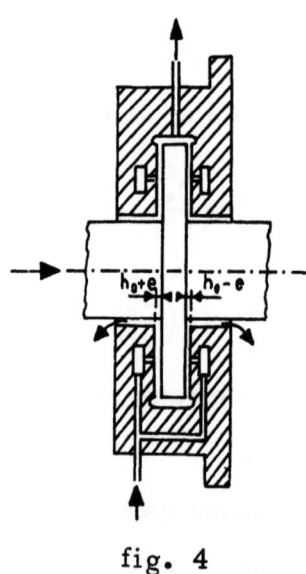

fig. 4

Applications to Statical Gas Bearings

One recognizes the gas supply through a row of orifices and the two gaps of different thickness. There ensues an excentricity e brought about by the load on the bearings. The gas exit takes place both at the outside radius r_a and the inside radius r_i.

We shall consider bearings with lòw overpressures as they come in question for gimbal bearings for gyroscopes. In this case we can neglect the inertial terms and we can make use of the approximation (41) for the flow through the orifices. In the bearing itself we can calculate with a laminar flow. The gap shows a rotational symmetry and the thickness of the gap is produced according to the axial force P. With the bearing working in two directions the total thickness of the gaps is $2h_0$. For the thickness of the left gap h_l and for the thickness of the right one h_r respectively, we have the equations:

$$h_l = h_0(1 + \varepsilon) \qquad (42)$$

$$h_r = h_0(1 - \varepsilon) \qquad (43)$$

with
$$\varepsilon = \frac{e}{h} \qquad (44)$$

e is the excentricity of the gap depending upon the load. The orifices for the gas supply shall be distributed equally on a graduated circle with the radius r_E (figure 5).

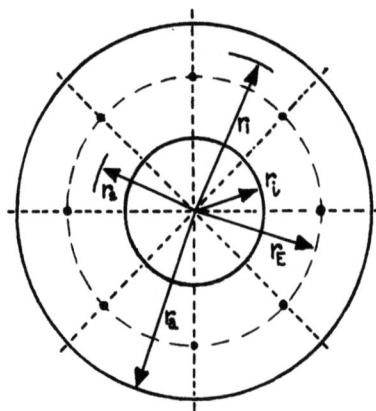

Figure 5.

In the outward field of the flow the exit of the gas takes place at the circle with the radius r_a, in the inward field at the circle with the radius r_i. Therefore the gas coming from the orifices branches off in two directions of flow. The gas supply through the orifices are substituted by the equal source distribution along the circumference of the circle with the radius r_E. Under the above assumption equation (5) can be applied for the flow through the gap. If r_1 is the variable radius in the outward field, r_2 in the inward one, p_1, p_2 the respective pressures, we have:

$$(45) \quad \frac{dp_1}{dr_1} = -12\eta \frac{\bar{u}_1}{h^2}$$

$$(46) \quad \frac{dp_2}{dr_2} = +12\eta \frac{\bar{u}_2}{h^2}.$$

The change of sign results from the change of the direction of the flow. The equation of continuity reads for the two fields,

Application to Statical Gas Bearings

in the case of rotational symmetry, analogous to eq. (7):

$$r_1 p_1 \bar{u}_1 = K_1, \qquad (47)$$

$$r_2 p_2 \bar{u}_2 = K_2. \qquad (48)$$

The constants K_1 and K_2 depend on the supply of the gas. If we eliminate $\bar{u}_1$ and $\bar{u}_2$ respectively from (45) and (47) and from (46) and (48) respectively we get after integrating:

$$p_1^2 - p_{cr}^2 = \frac{24\eta K_1}{h^2} \ln\frac{r_a}{r_1}, \qquad (49)$$

$$p_2^2 - p_{cr}^2 = \frac{24\eta K_2}{h^2} \ln\frac{r_2}{r_i}. \qquad (50)$$

In the above equations p_{ex} is the outward pressure and the following boundary conditions hold true: $r_1 = r_a \ldots p_1 = p_{ex}$ and $r_2 = r_i \ldots p_2 = p_{ex}$. If $\frac{dm_1}{dt}$ and $\frac{dm_2}{dt}$ mean the shares of the mass-flow of the gas that flow into the outward and inward field per orifice, and if n orifices are arranged on each side, then we have, because of the conservation of mass:

$$n\frac{dm_1}{dt} = 2\pi r_1 h \varrho_1 \bar{u}_1,$$

$$n\frac{dm_2}{dt} = 2\pi r_2 h \varrho_2 \bar{u}_2.$$

As the flow is isothermal we have, because of the equation of state, the temperature T_0 remaining constant, if we take

(47) and (48) into account:

$$n\frac{dm_1}{dt} = \frac{2\pi h}{RT_0} K_1 \quad \text{and} \quad n\frac{dm_2}{dt} = \frac{2\pi h}{RT_0} K_2.$$

The total flow of mass $n\frac{dm}{dt}$ through one ring of orifices is therefore:

(51)
$$n\frac{dm}{dt} = \frac{2\pi h}{RT_0} (K_1 + K_2).$$

If we neglect the radial extension of the field of orifices, then for the radius: $r_1 = r_2 = r_E$, the pressure must be: $p_1 = p_2 = p_E$, where p_E is the substitute pressure along the circumference of source distribution in the ring of orifices. There results therefore, from (49) and (50):

(52)
$$\frac{K_1}{K_2} = \frac{\ln \frac{r_E}{r_i}}{\ln \frac{r_a}{r_E}}.$$

There results from (51) and (52):

(53)
$$K_1 = \frac{RT_0}{2\pi h} n \frac{dm}{dt} \frac{\ln \frac{r_E}{r_i}}{\ln \frac{r_a}{r_i}},$$

(54)
$$K_2 = \frac{RT_0}{2\pi h} n \frac{dm}{dt} \frac{\ln \frac{r_a}{r_E}}{\ln \frac{r_a}{r_i}}.$$

The quantity $\frac{dm}{dt}$ can be calculated according to eq. (41). Instead of p_0 and ϱ_0 we write now p_s and ϱ_s (feeding pressure

Application to Statical Gas Bearings

and feeding density). When making use of the equation of state:

$$\frac{p_s}{\varrho_s} = RT_0 \quad \text{we get} \quad \frac{dm}{dt} = \alpha A_E \sqrt{\frac{2p_s(p_s - p_E)}{RT_0}} \quad (55)$$

where T_0 stands for the isothermal temperature. The pressure p_E produced in the ring of orifices results from (49), (53) and (55) for $r_1 = r_E$ or, which gives the same result, from (50), (54) and (55) for $r_2 = r_E$. For p_E we get the equation:

$$p_E^2 - p_{ex}^2 = \frac{12\eta\sqrt{RT_0}}{\pi h^3} n\alpha A_E \sqrt{2p_s(p_s - p_E)} \frac{\ln\frac{r_a}{r_E} \ln\frac{r_E}{r_i}}{\ln\frac{r_a}{r_i}}. \quad (56)$$

This is an equation of the fourth degree for p_E which shall not be solved here explicitely.

If $p_{E\ell}$ respectively p_{Er} denotes the pressures in the ring of orifices of the left and the right bearings respectively and p_{E0} the respective pressure in the unloaded bearing ($h = h_0$), there results from (56):

$$\frac{p_{E\ell}^2 - p_{ex}^2}{p_{E0}^2 - p_{ex}^2} = \frac{h_0^3}{h_\ell^3}\sqrt{\frac{p_s - p_{E\ell}}{p_s - p_{E0}}} \quad (57)$$

p_{E0} results from (56) for $h = h_0$.

From (57) and (42) we get for $p_{E\ell}$ if one restricts oneself to linear terms in ε:

$$p_{E\ell} = p_{E0}\left[1 - 6\frac{(p_s - p_{E0})(p_{E0}^2 - p_{ex}^2)}{p_{E0}(4p_s p_{E0} - 3p_{E0}^2 - p_{ex}^2)}\varepsilon\right]. \quad (58)$$

If we substitute in the above equation ε by $-\varepsilon$ we get for p_{Er}

$$(59) \quad p_{Er} = p_{E0}\left[1 + 6\frac{(p_s - p_{E0})(p_{E0}^2 - p_{ex}^2)}{p_{E0}(4p_s p_{E0} - 3p_{E0}^2 - p_{ex}^2)}\varepsilon\right].$$

For $\varepsilon > 0$ we have $p_{Er} > p_{E0} > p_{E\ell}$ i.e. the smaller pressure in the orifice belongs to the bigger gap and vice versa.

If we call K_1 and K_{10} respectively the values of K_1 for $h = h_\ell$ and for $h = h_0$, respectively, we get from (53) and (55):

$$(60) \quad \frac{K_{1\ell}}{K_{10}} = \frac{h_0}{h_\ell}\sqrt{\frac{p_s - p_{E\ell}}{p_s - p_{E0}}}.$$

From (58) and (42) there follows, if we expand till to the linear terms in ε:

$$(61) \quad K_{1\ell} = K_{10}\left[1 - \frac{4p_s p_{E0} - 3p_{E0}^2 + 2p_{ex}^2}{4p_s p_{E0} - 3p_{E0}^2 - p_{ex}^2}\varepsilon\right]$$

and for

$$(62) \quad K_{1r} = K_{10}\left[1 + \frac{4p_s p_{E0} - 3p_{E0}^2 + 2p_{ex}^2}{4p_s p_{E0} - 3p_{E0}^2 - p_{ex}^2}\varepsilon\right].$$

From (49) we get: $\dfrac{p_{1\ell}^2 - p_{ex}^2}{p_{10}^2 - p_{ex}^2} = \dfrac{h_0^2 K_{1\ell}}{h_\ell^2 K_{10}}$. From it we get using (42) and (61) expanding till to the terms of the first degree in ε:

$$(63) \quad p_{1\ell} = p_{10} - 6\left(p_{10} - \frac{p_{ex}^2}{p_{10}}\right)\cdot\frac{p_{E0}(p_s - p_{E0})}{4p_s p_{E0} - 3p_{E0}^2 - p_{ex}^2}\varepsilon$$

Application to Statical Gas Bearings

$$p_{1r} = p_{10} + 6\left(p_{10} - \frac{p_{ex}^2}{p_{10}}\right) \cdot \frac{p_{E0}(p_s - p_{E0})}{4p_s p_{E0} - 3p_{E0}^2 - p_{ex}^2} \varepsilon \quad . \quad (64)$$

The resulting pressure Δp_1 with both bearings cooperating gives:

$$\Delta p_1 = p_{1r} - p_{1\ell} = 12\left(p_{10} - \frac{p_{ex}^2}{p_{10}}\right) \cdot \frac{p_{E0}(p_s - p_{E0})}{4p_s p_{E0} - 3p_{E0}^2 - p_{ex}^2} \varepsilon \quad (65)$$

and analogous:

$$\Delta p_2 = p_{2r} - p_{2\ell} = 12\left(p_{20} - \frac{p_{ex}^2}{p_{20}}\right) \cdot \frac{p_{E0}(p_s - p_{E0})}{4p_s p_{E0} - 3p_{E0}^2 - p_{ex}^2} \varepsilon \quad . \quad (66)$$

p_{10} and p_{20} can be represented by (49) and (50). Expanding the square roots according to the binomial theorem we get, if we restrict ourselves to linear terms in K_{10} and K_{20}, in accordance with the above degree of approximation:

$$p_{10} = p_{ex}\left[1 + \frac{12\eta K_{10}}{h_0^2 p_{ex}^2} \ln \frac{r_a}{r_1}\right] \quad (67)$$

$$p_{20} = p_{ex}\left[1 + \frac{12\eta K_{20}}{h_0^2 p_{ex}^2} \ln \frac{r_2}{r_i}\right] \quad . \quad (68)$$

Using the equations (53), (54), (55) and (56) we get from (67) and (68):

$$p_{10} - \frac{p_{ex}^2}{p_{10}} = \frac{p_{E0}^2 - p_{ex}^2}{p_{ex}} \cdot \frac{\ln \frac{r_a}{r_1}}{\ln \frac{r_a}{r_E}} \quad (69)$$

$$p_{20} - \frac{p_{ex}^2}{p_{20}} = \frac{p_{E0}^2 - p_{ex}^2}{p_{ex}} \cdot \frac{\ln \frac{r_2}{r_i}}{\ln \frac{r_E}{r_i}} \quad . \quad (70)$$

If P_A is the axial force operating on the double bearing we get:

$$(71) \quad P_A = 2\pi \int_{r_E}^{r_a} r_1 \Delta p_1 dr_1 + 2\pi \int_{r_i}^{r_E} r_2 \Delta p_2 dr_2 .$$

As axial stiffness for small displacements we can define:

$$(72) \quad k_A = \frac{P_a}{e} .$$

In it e means $h_0 \epsilon$.

Carrying out the integration (71) we get, when using the equations (53), (54), (65), (66), (69), (70), (71) and (72):

$$(73) \quad k_A = \frac{6\pi}{h_0 \ln\frac{r_a}{r_E} \ln\frac{r_E}{r_i}} \left[r_a^2 \ln\frac{r_E}{r_i} - r_E^2 \ln\frac{r_a}{r_i} + r_i^2 \ln\frac{r_a}{r_E} \right] \frac{p_{EO}}{p_{ex}} \frac{(p_{EO}^2 - p_{ex}^2)(p_s - p_{EO})}{4 p_s p_{EO} - 3 p_{EO}^2 - p_{ex}^2} .$$

Eq. (73) contains all dependences of the axial stiffness upon the constructive data and upon the pressures. If we solve eq. (56) with respect to p_E substituting h_0 for h we get p_{EO} which we find in eq. (73). We recognize that the decrease of pressure $(p_s - p_{EO})$ in the flow in the orifices in case of the centred position is decisive for the stiffness. If we do not allow any decrease of pressure in the orifices the bearing will lose its stiffness. To a certain degree we can optimize the construction and the operating data. Apparently it is useful to make a minimum of the power of the compressor that supplies the double bearing, where we can use both $p_s - p_{EO}$ and r_E as optimizing parameter. However we shall not go into details here.

Application to Statical Gas Bearings

We now want to discuss the journal bearing. (fig. 6).

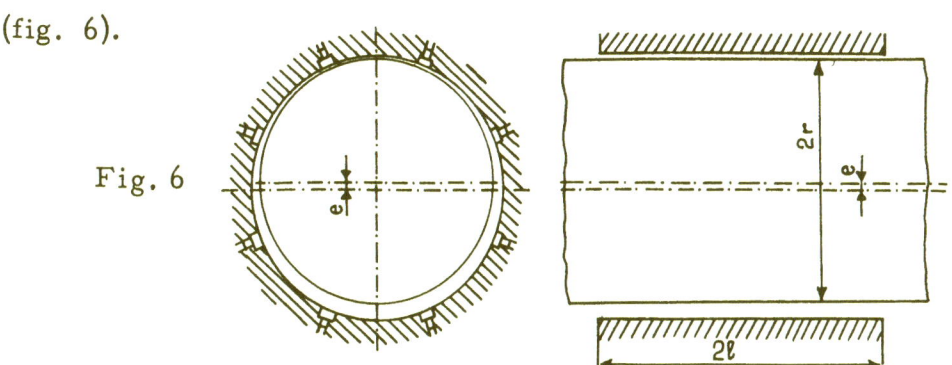

Fig. 6

The gas supply takes place over a ring of orifices which is assumed to lie in the plain of symmetry. For small excentricities e between the centres of the pivot and the bearing-housing we have the following relation for the thickness h of the gas-gap according to figure (7):

$$h = h_0(1 - \varepsilon \cos\varphi) \qquad (74)$$
$$\text{with} \quad \varepsilon = \frac{e}{r} \qquad (75)$$

where r is the radius of the pivot (Fig. 7). If we neglect the inertial force and the friction due to compression we get eq. (5) for both directions of the co-ordinates x (axial direction) and y (circumferential direction):

$$\frac{\partial p}{\partial x} = -12\eta \frac{\bar{u}_x}{h^2} \qquad (76)$$

$$\frac{\partial p}{\partial y} = -12\eta \frac{\bar{u}_y}{h^2} \qquad (77)$$

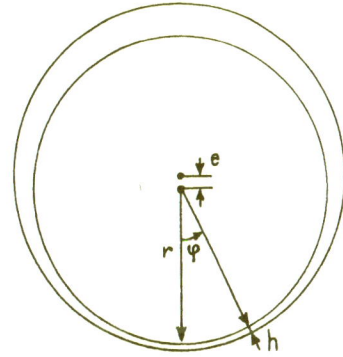

Fig. 7

where $\bar{u}_x$ and $\bar{u}_y$ are the flow-velocities in the x-and y-direction averaged over the thickness of the gap. (Unrolling of the circumference of a cylinder onto a plain).

The equation of continuity reads:

(78) $$\frac{\partial}{\partial x}(\varrho h u_x) + \frac{\partial}{\partial y}(\varrho h u_y) = 0$$

or, if we substitute for $\varrho = \frac{p}{RT_0}$ according to the equation of continuity and if we assume isothermal change of state, taking into account (76) and (77):

$$\frac{\partial}{\partial x}\left(h^3 p \frac{\partial p}{\partial x}\right) + \frac{\partial}{\partial y}\left(h^3 p \frac{\partial p}{\partial y}\right) = 0.$$

If we substitute

(79) $$y = r\varphi$$

we get : (see figure 7)

with

(80) $$P = p^2.$$

Taking into account eq. (74) we finally get :

$$h^3\left(\frac{\partial^2 P}{\partial x^2} + \frac{1}{r^2}\frac{\partial^2 P}{\partial \varphi^2}\right) + \frac{3h^2}{r^2}h_0 \varepsilon \sin\varphi \frac{\partial P}{\partial \varphi} = 0.$$

As for $\varepsilon = 0$ does not depend upon φ because of the symmetry, $\frac{\partial P}{\partial \varphi}$ has the order of magnitude ε. Restricting our-

Application to Statical Gas Bearings

selves to linear terms in ε we get:

$$\frac{\partial^2 P}{\partial x^2} + \frac{1}{r^2}\frac{\partial^2 P}{\partial \varphi^2} = 0 \ . \tag{81}$$

If ℓ is half of the length of the bearing there holds the boundary condition:

$$x = \ell \ldots \ldots P = P_{ex} = P_{ex}^2 \tag{82}$$

if P_{ex} means the constant ambient pressure.

The general solution of (81) fulfilling the boundary condition (82) is:

$$P = \sum_{j=1}^{\infty}\left[B_j \, sh\frac{j(\ell - x)}{r}\cos j\varphi\right] + C(\ell - x) + P_{ex} \ . \tag{83}$$

In the above and in the following equations the hyperbolic functions are indicated by sh and ch. A second boundary condition is given by the supply condition for the ring of orifices. If we substitute the discrete orifices by a linear source whose intensity is a steady function of φ, then we have, according to the equation of continuity, as the same mass-flow takes place on both sides:

$$n\frac{dm}{dt}\frac{d\varphi}{2\pi} = 2\rho h \bar{u}_x r \, d\varphi \ . \qquad \text{(compare figure 8)}$$

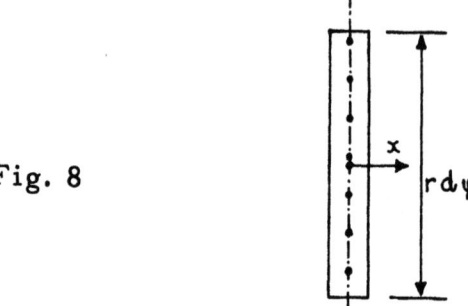

Fig. 8

Taking into account eq. (76), the equation of state and the equation (80) the above yields:

(84) $$n\frac{dm}{dt} = -\frac{\pi r}{6RT_0 \eta} h^3 \frac{\partial P}{\partial x}\bigg|_{x=0}$$

Eq. (84) has to be satisfied for $x = 0$ for all φ.

We restrict ourselves to small excentricities and cancel all terms in ϵ that are of a higher than the first degree. Accordingly we state in (83):

(85) $$B_j = B_{j0} + \epsilon B_{j1}$$

and

(86) $$C = C_0 + \epsilon C_1 .$$

As for $\epsilon = 0$ becomes independent from φ we follow from (83):

(87) $$B_{j0} = 0 \quad (j = 1,2,3.....) .$$

With (83), (85), (86) and (87), the condition of

supply (84) reads :

$$n\frac{dm}{dt} = \frac{\pi r}{6RT_0\eta}h^3\left\{C_0 + \varepsilon\left[C_1 + \sum_{j=1}^{\infty}(B_{j1}\frac{j}{r}ch(\frac{jl}{r})\cos j\varphi)\right]\right\}. \quad (88)$$

If P_{E0} is the value of P for $x = 0$ in case of the bearing being unloaded ($\varepsilon = 0$) we have from (83) taking account of (85), (86) and (87) :

$$P_{E0} = C_0 l + P_{ex} . \quad (89)$$

With the above there results from (83), (85), (86) and (87) :

$$P = P_{E0} - (P_{E0} - P_{ex})\frac{x}{l} + \varepsilon\sum_{j=1}^{\infty}(B_{j1} sh\frac{j(l-x)}{r}\cos j\varphi) . \quad (90)$$

For $\frac{dm}{dt}$ we substitute the value according to eq. (55). With the above there results from (88), taking account of (89) and (80) :

$$\frac{n\alpha A_E}{\sqrt{RT_0}}\sqrt{2p_S(p_S-p_E)} =$$

$$= \frac{\pi r}{6RT_0\eta}h^3\left\{\frac{p_{E0}^2 - p_{ex}^2}{l} + \varepsilon\left[C_1 + \sum_{j=1}^{\infty}B_{j1}\frac{j}{r}ch(\frac{jl}{r})\cos j\varphi\right]\right\}. (91)$$

Eq. (90) yields for $x = 0$ the value $P = P_E$ as follows :

$$P_E = p_E^2 = p_{E0}^2 + \varepsilon\sum_{j=1}^{\infty}(B_{j1} sh(\frac{jl}{r})\cos j\varphi) . \quad (92)$$

Using eq. (91) for the unloaded bearing ($\varepsilon = 0$) we get the following relation for p_{E0}:

$$(93) \quad p_{E0}^2 - p_{ex}^2 = \frac{6n\alpha A_E \eta \sqrt{RT_0}}{\pi h_0^3} \frac{\ell}{r} \sqrt{2p_s(p_s - p_{E0})} .$$

If we expand the function p_E till the linear terms in $\mathcal{E}$ using eq. (92) substituting this value in eq. (91) and expanding again till to the linear terms in $\mathcal{E}$ we can carry out a comparison of coefficients for the left and the right Fourier-series. This yields the following relations for the unknown constants C_1 and B_{j1} using eq. (93):

$$(94) \quad C_1 = 0 ,$$

$$(95) \quad B_{11} = \frac{12 p_{E0}(p_{E0}^2 - p_{ex}^2)(p_s - p_{E0})}{(p_{E0}^2 - p_{ex}^2)\,\mathrm{sh}\frac{\ell}{r} + 4\frac{\ell}{r} p_{E0}(p_s - p_{E0})\,\mathrm{ch}\frac{\ell}{r}} ,$$

$$(96) \quad B_{j1} = 0 \quad (j = 2, 3, 4 \ldots) .$$

Substituting these values in eq. (90) we get for $p = \sqrt{P}$ excluding the square terms and higher terms in $\mathcal{E}$:

$$(97)\, p = \sqrt{p_{E0}^2 - (p_{E0}^2 - p_{ex}^2)\frac{x}{\ell}} + \mathcal{E} S \frac{\mathrm{sh}\left(\frac{\ell - x}{r}\right)}{\sqrt{p_{E0}^2 - (p_{E0}^2 - p_{ex}^2)\frac{x}{\ell}}} \cos\varphi$$

with the abbreviation :

$$(98) \quad S = \frac{6 p_{E0}(p_{E0}^2 - p_{ex}^2)(p_s - p_{E0})}{(p_{E0}^2 - p_{ex}^2)\,\mathrm{sh}\frac{\ell}{r} + 4\frac{\ell}{r} p_{E0}(p_s p_{E0})\,\mathrm{ch}\frac{\ell}{r}} .$$

Application to Statical Gas-Bearings

The supporting force K_a directed perpendicularly to the pivot axis results from :

$$K_a = 2\int_0^\ell dx \int_0^{2\pi} p\cos\varphi\, r\, d\varphi .$$

The respective perpendicular stiffness : $k = \dfrac{K_a}{e} = \dfrac{K_a}{\varepsilon h_0}$ receives the form :

$$k_a = 2\pi \frac{r}{h_0} S \int_0^\ell \frac{\operatorname{sh}\left(\frac{\ell-x}{r}\right) dx}{\sqrt{p_{E0}^2 - (p_{E0}^2 - p_a^2)\frac{x}{\ell}}} .$$

According to our above approximation it is sufficient to expand the root in the denominator till to the linear terms in x. This yields :

$$k_a = \frac{r^2 \pi}{h_0 p_{E0}^3}\left[2p_{E0}^2(\operatorname{ch}\tfrac{\ell}{r} - 1) - (\tfrac{r}{\ell}\operatorname{sh}\tfrac{\ell}{r} - 1)(p_{E0}^2 - p_{ex}^2)\right] S . \quad (99)$$

One recognizes from eq. (99) in connection with eq. (98) that k_a approaches nought with the decrease of pressure in the orifice. Again we can optimize p_{E0} to get the minimum of power of the compressors; in the same way we can optimize the fraction : $\dfrac{\ell}{r}$. Then eq. (93) shows how this most favourable pressure p_{E0} can be realized in the ring of orifices by the lay-out of the cross-section of the orifice.

4. Self-Supporting Bearings for the Spin-Axis.

Owing to the high speed of rotations one can, to support spin axis, use self-supporting gas-bearings the support of which is effected by the speed of rotations. In this case the axis is fixed while the housing of the bearing firmly connected with the gyro-rotor rotates. To enhance the effect of supporting spiral grooves are often milled into the axis. Figure 9 and Figure 10 show a herring bone journal bearing

Fig. 9   Fig. 10

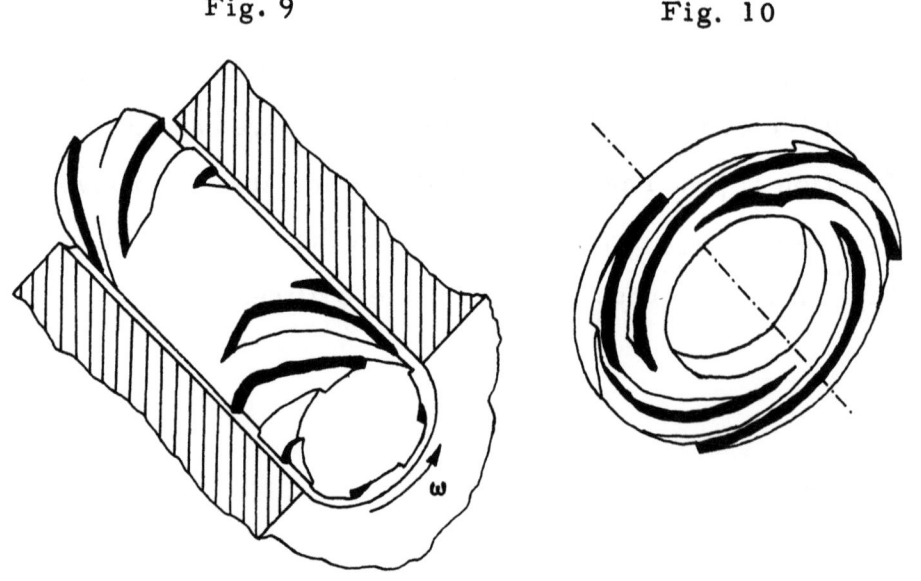

and a thrust bearing of this kind. In order not to make the calculations too complicated we shall only discuss the smooth journal bearings. As we also want to treat instabilities we must assume the flow in the gap as not stationary. Neglecting the inertial terms and the friction owing to the compression we follow from the vectorial Stokes-Navier equation (2):

$$\nabla p = \eta \Delta \overset{a}{u} = \eta \frac{\partial^2 \overset{a}{u}}{\partial z^2} . \qquad (100)$$

Integrating twice over z with the boundary condition $z = 0 \ldots \overset{a}{u} = 0$ and $z = h \ldots u = \mathscr{w}$, we get:

$$\overset{a}{u} = \frac{\nabla p}{2\eta}(z^2 - hz) + \mathscr{w}\frac{z}{h}$$

$\mathscr{w}$ is the vectorial velocity of the point in question of the housing. If $\overset{a}{u}$ is the vectorial velocity of flow averaged over the gap we get:

$$\overset{a}{u} = -\frac{h^2}{12\eta}\nabla p + \frac{1}{2}\mathscr{w} . \qquad (101)$$

The equation of continuity for the flow in the gap reads:

$$\nabla(\rho h \overset{a}{u}) + \frac{\partial(\rho h)}{\partial t} = 0 .$$

In case of isothermal change of state we can substitute ρ by p. Taking into account eq. (101) the equation of continuity gets the form :

(102) $$\nabla(\frac{h^3}{12\eta} p\nabla p - \frac{hp}{2}\omega) = \frac{\partial(ph)}{\partial t}.$$

This is the isothermal Reynolds' equation. Usually a dimensionless representation is preferred. If we write :

(103) $x = r_0 x^*;\ y = r_0 y^*;\ h = h_0 h^*;\ p = P_{ex} p^*;\ \omega = r_0 \omega w^*;\ t = \frac{1}{\nu} t^*$

then we have :

(104) $$\nabla = n_x \frac{\partial}{\partial x} + n_y \frac{\partial}{\partial y} = \frac{1}{r_0} \nabla^*$$

ω and ν are angular velocities, n_x, n_y are unit vectors.
If we substitute (103) and (104) in (102) we get :

(105) $$\nabla^* \cdot (h^{*3} p^* \nabla^* p^* - \Lambda h^* p^* w^*) = \sigma \frac{\partial(h^* p^*)}{\partial t^*}$$

with :

(106) $$\Lambda = \frac{6\eta \omega r_0^2}{P_{ex} h_0^2}$$

and :

(107) $$\sigma = \frac{12\eta \nu r_0^2}{P_{ex} h_0^2} = 2(\frac{\nu}{\omega})\Lambda$$

Λ is called compressibility number, σ is a disturbance that occurs in case of instability.

We shall show the circumstances in a simple

Self-Supporting Bearings for the Spin Axis

example. We are thinking of a journal bearing according to figure 11. (see page 43). In a centred position the gas gap is h_0. If the rotation is stationary the load $\vec{K}$ being constant, there will be developed a constant angle Φ between the load vector and the displacement vector in the radial plain.

If $\vec{\kappa}$ is the displacement vector, then there results a disturbance moment operation on the gyroscope with the vector $\vec{\kappa} \times \vec{K}$, as $\vec{\kappa}$ is not parallel with $\vec{K}$. This disturbance moment is so small in general that its effect can be neglected. However, nonsteady disturbance motions are much more critical, therefore we shall discuss them in detail. We can assume approximately that the gyroscope is unloaded, and that the housing together with the rotor carry out a so-called whirl motion. This is a steady rotation of the centre of mass of the rotor around the centre of the axis with the excentricity e_r and the angular velocity $\dot{\Phi}$ being constant. The spin-axis in this case remains parallel to itself. The angle measured from a fixed direction to the direction of the displacement vector grows proportionally to the time:

$$\dot{\Phi} = \nu t. \tag{108}$$

In case of small excentricity we can write for the gap thickness h:

(109) $$h = h_0 + \text{Real}\left\{e_r e^{i(\theta - vt)}\right\}.$$

We state for the gas pressure in the gap:

(110) $$p = p_{ex} + \text{Real}\left\{e_r p_r e^{i(\theta - vt)}\right\}.$$

In it "Real" means the "real part of".

Making the quantities dimensionless according to eq. (103) we get:

(111) $$h^* = 1 + \text{Real}\left\{\varepsilon_r e^{i(\theta - t^*)}\right\}$$

(112) $$p^* = 1 + \text{Real}\left\{\varepsilon_r p_r^* e^{i(\theta - t^*)}\right\},$$

with:

(113) $$\varepsilon_r = \frac{e_r}{h_0}$$

and:

(114) $$p_r^* = \frac{p_r}{p_{ex}}.$$

As we consider a journal bearing where the flow takes place in a plane, small boundary disturbances excepted, we can restrict ourselves to the plane problem and we get the angle θ as the only coordinate. According to (104) we read :

$$\nabla^* = n_\theta \frac{\partial}{\partial \theta} . \qquad (115)$$

Further on we write : $\wp = n_\theta r_0 \omega$ (116)

if n_θ is the unit vector in the direction of the coordinate and ω the angular velocity of the gyroscope wheel. According to eq. (103) we get : $\wp^* = n_\theta$. (117)

It is useful to take a new variable : $\underline{\theta} = \theta - t^*$ (118)

the meaning of which is to be seen in figure 11.

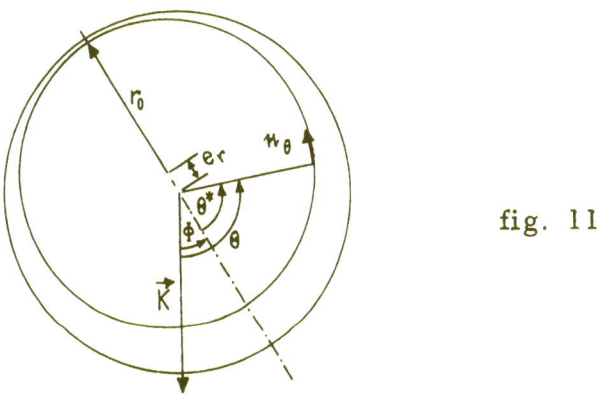

fig. 11

Then we have for the partial differential operators :

$$\frac{\partial}{\partial \theta}\bigg|_{t^*} = \frac{\partial}{\partial \underline{\theta}^*}\bigg|_{t^*} \; ; \; \frac{\partial}{\partial t^*}\bigg|_\theta = \frac{\partial}{\partial t^*}\bigg|_{\underline{\theta}^*} - \frac{\partial}{\partial \underline{\theta}^*}\bigg|_{t^*} . \qquad (119)$$

With the above eq. (105) gets the form :

$$n_\theta \frac{\partial}{\partial \underline{\theta}^*}\left[h^{*3} p^* n_\theta \frac{\partial p^*}{\partial \underline{\theta}^*} - \Lambda n_\theta h^* p^*\right] = \sigma\left[\frac{\partial}{\partial t^*}(h^* p^*) - \frac{\partial}{\partial \underline{\theta}^*}(h^* p^*)\right] .$$

As, according to the statements eq. (111) and (112) taking into account eq. (118), h^* and p^* only depend upon Θ^* and not upon t^* we get :

(120) $$\Lambda_\Theta \frac{\partial}{\partial \Theta^*}\left(h^{*3} p^* \Lambda_\Theta \frac{\partial p^*}{\partial \Theta^*}\right) - \Lambda^* \frac{\partial}{\partial \Theta^*}(h^* p^*) = 0$$

with :

(121) $$\Lambda^* = \Lambda - \sigma = \frac{6\eta r_0^2}{P_{ex} h_0^2}(\omega - 2\nu)$$

if we take into account eq. (106) and (107).

If we substitute the statements (111) and (112) in eq. (120), taking into account only the linear terms in ε_r we get :

$$\text{Real}\left\{\varepsilon_r e^{i\Theta^*}\left[p_r^* + i\Lambda^*(p_r^* + 1)\right]\right\} = 0.$$

As this equation must hold true for all Θ^* we get :

(122) $$p_r^* = -\frac{\Lambda^*}{1+\Lambda^{*2}}(\Lambda^* + i) .$$

The pressure p^* becomes, according to eq. (112) and (118) :

(123) $$p^* = 1 + \varepsilon_r \frac{\Lambda^*}{1+\Lambda^{*2}}(-\Lambda^* \cos\Theta^* + \sin\Theta^*) .$$

Following (103) and (113) we have :

(124) $$p = P_{ex}\left[1 + \frac{e_r}{h_0}\frac{\Lambda^*}{1+\Lambda^{*2}}(-\Lambda^* \cos\Theta^* + \sin\Theta^*)\right] .$$

Self-Supporting Bearings for the Spin Axis 45

We resolve the resulting pressure on the housing, its length being ℓ into the components $K_{\parallel}$ in the direction e_r, and $K_{\perp}$ in the direction perpendicular to the first one (positive anticlockwise).

Then we get (see figure 11):

$$K_{\parallel} = r_0 \ell \int_0^{2\pi} p \cos\Theta^* d\Theta^* \quad \text{and} \quad K_{\perp} = r_0 \ell \int_0^{2\pi} p \sin\Theta^* d\Theta^* .$$

We get the stiffnesses if we divide by $-e_r$. According to (124) we have:

$$k_{\parallel} = \frac{r_0 \pi \ell p_{ex}}{h_0} \frac{\Lambda^{*2}}{1 + \Lambda^{*2}} \qquad (125)$$

$$k_{\perp} = -\frac{r_0 \pi \ell p_{ex}}{h_0} \frac{\Lambda^*}{1 + \Lambda^{*2}} . \qquad (126)$$

There are more complicated relations with grooved bearings. We will not go into details.

We are going to treat now the self-excited instabilities, the stiffnesses $k_{\perp}$ and $k_{\parallel}$ being assumed as already known. Their dependence on v and ω needs not have the simple form of eq. (125) and (126) with (121) in more complicated cases. Then the spin of gyroscopes is assumed as constant, as well as the angular velocity ω of the rotor.

If x and y mean the displacements in two fixed directions perpendicular to the spin axis, and K_x and K_y

mean components of the force in the direction x and y operating on the housing owing to the displacements, then we can set down the following relation in the matrix notation for small displacements owing to the difference of direction between the resulting force and the resulting displacement :

$$(127) \qquad \begin{bmatrix} K_x \\ K_y \end{bmatrix} = - \begin{bmatrix} b_{xx} & b_{xy} \\ b_{yx} & b_{yy} \end{bmatrix} \begin{bmatrix} x \\ y \end{bmatrix}.$$

The sign minus has been inserted on the right side as the force and the displacement have different signs. As we do not assume any steady load of the rotor we have no distinguished direction, eq. (127) must be invariant respective to the rotation of the coordinate system. If we substitute accordingly $x \to y$ and $x \to -y$ then there becomes also $K_x \to K_y$ and $K_y \to -K_x$. There follows for the terms of the matrix :

$$(128) \qquad b_{xx} = b_{yy} = b_{\|}$$

$$(129) \qquad b_{xy} = -b_{yx} = b_{\perp}$$

$b_{\|}$ and $b_{\perp}$ are new abbreviations.

The centre of mass law applied to the movement of the rotor reads :

$$(130) \qquad m \frac{d^2}{dt^2} \begin{bmatrix} x \\ y \end{bmatrix} = \begin{bmatrix} K_x \\ K_y \end{bmatrix}.$$

Self-Supporting Bearings for the Spin-Axis 47

In it m is the mass of the rotor.

From (127) and (130) there results, taking into account (128) and (129):

$$\begin{bmatrix} b_{\parallel} + m\dfrac{d^2}{dt^2} & b_{\perp} \\ -b_{\perp} & b_{\parallel} + m\dfrac{d^2}{dt^2} \end{bmatrix} \begin{bmatrix} x \\ y \end{bmatrix} = 0 . \quad (131)$$

Eq. (131) holds true in case of the external load of the rotor being absent.

The relation between the values $b_{\parallel}$ and $b_{\perp}$ and the stiffnesses will be stated later on. To solve eq. (131) we state:

$$\begin{bmatrix} x \\ y \end{bmatrix} = \begin{bmatrix} x_0 \\ y_0 \end{bmatrix} \exp[(\lambda + i\nu)t] . \quad (132)$$

Thus, eq. (131) reads:

$$\begin{bmatrix} b_{\parallel} + m(\lambda + i\nu)^2 & b_{\perp} \\ -b_{\perp} & b_{\parallel} + m(\lambda + i\nu)^2 \end{bmatrix} \begin{bmatrix} x_0 \\ y_0 \end{bmatrix} = 0 . \quad (133)$$

We get non-trivial solutions by the vanishing of the determinant belonging to the matrix in eq. (133). This yields the equation:

$$\left[b_{\parallel} + m(\lambda + i\nu)^2 \right]^2 + b_{\perp}^2 = 0 \quad (134)$$

respectively:

(135) $$b_{\|} \mp ib_{\perp} + m(\lambda + i\nu)^2 = 0.$$

This is the characteristic equation for the radial motion.

In general $b_{\|}$ and $b_{\perp}$ will be complex numbers. Therefore we write:

(136) $$\begin{aligned} b_{\|} &= u_{\|} + i\nu_{\|} \\ b_{\perp} &= u_{\perp} + i\nu_{\perp} . \end{aligned}$$

The stability threshold is distinguished by $\lambda = 0$. If we write $\lambda = 0$ in eq. (135) using the equations (136) we get after separating the real part from the imaginary part:

(137) $$(u_{\|} \pm \nu_{\perp})|_{\nu_{cr}} - m_{cr} \nu_{cr}^2 = 0$$

(138) $$(\nu_{\|} \mp u_{\perp})|_{\nu_{cr}} = 0.$$

In the above the values of m and ν were written: m_{cr} and ν_{cr}. In general we get two critical values of m_{cr} and ν_{cr}.

To distinguish the stable from the unstable field we also need the deviations from the values $\lambda + i\nu$ in the surroundings of m_{cr} we write

(139) $$\begin{aligned} m &= m_{cr} + m' . \\ \lambda + i\nu &= i\nu_{cr} + (\lambda' + i\nu') . \end{aligned}$$

Self-Supporting Bearing for the Spin-Axis

If we substitute the values of eq. (130) in eq. (135) we get, taking into account eq. (136), expanding into a Taylor' series till to the linear terms in $\lambda' + i\nu'$:

$$\left[-i\frac{\partial(u_{\|} \pm v_{\perp})}{\partial \nu} + 2i(u_{\|} \pm v_{\perp}) + \frac{\partial(v_{\|} \mp u_{\perp})}{\partial \nu}\right]\Bigg|_{\nu_{cr}} (\lambda' + i\nu') = m'v_{cr}^2$$

By rewriting the above we get:

$$\lambda' + i\nu' = \frac{i v_{cr}^2 \cdot m'}{\left[\nu^2 \frac{\partial}{\partial \nu}\left(\frac{u_{\|} \pm v_{\perp}}{\nu^2}\right) + i\frac{\partial}{\partial \nu}(v_{\|} \mp u_{\perp})\right]\Bigg|_{\nu_{cr}}} \quad (140)$$

The real part of the right term yields:

$$\lambda' = \left\{\frac{\nu^2 \frac{\partial}{\partial \nu}(v_{\|} \mp v_{\perp})m'}{\left[\nu^2 \frac{\partial}{\partial \nu}\left(\frac{u_{\|} \pm v_{\perp}}{\nu^2}\right)\right]^2 + \left[\frac{\partial}{\partial \nu}(v_{\|} \mp v_{\perp})\right]^2}\right\}\Bigg|_{\nu_{cr}} \quad (141)$$

In order to express the quantities $b_{\|}$ and $b_{\perp}$ respectively $u_{\|}$, $u_{\perp}$ and $v_{\|}$, $v_{\perp}$ by means of the stiffnesses $k_{\|}$ and $k_{\perp}$ we at first consider an anticlockwise steady-state whirl motion (i.e. in the same direction as the rotation of the gyro-rotor). The positive directions of the components $e_r k_{\|}$ and $e_r k_{\perp}$ of the forces transmitted from the housing on the gyro-rotor are shown in figure 12. The displacements belonging to this forward whirl motion f_x and f_y are on account of figure 12:

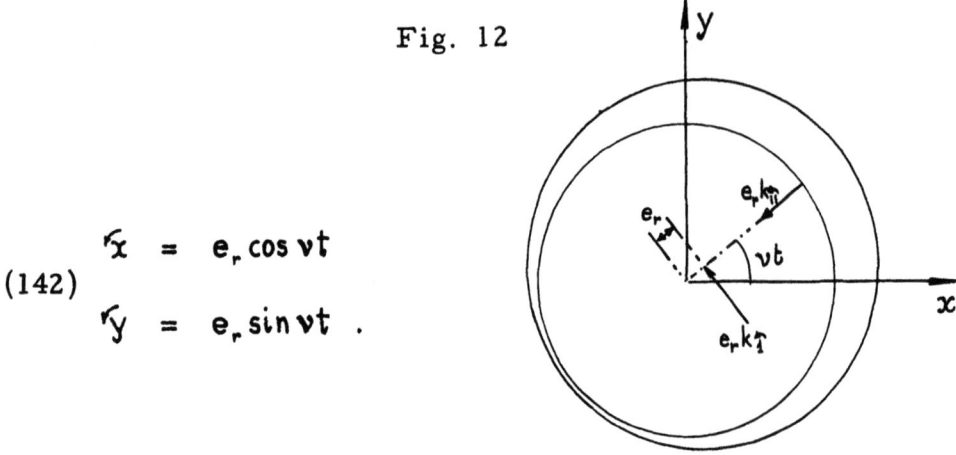

Fig. 12

(142)
$$\hat{x} = e_r \cos \nu t$$
$$\hat{y} = e_r \sin \nu t .$$

We get the components $K_{\hat{x}}$ and $K_{\hat{y}}$ in the x and y direction according to figure 12 :

(143)
$$K_{\hat{x}} = (-k_{\hat{r}\parallel} \cos \nu t + k_{\hat{r}\perp} \sin \nu t) e_r$$
$$K_{\hat{y}} = (-k_{\hat{r}\parallel} \sin \nu t + k_{\hat{r}\perp} \cos \nu t) e_r .$$

If the whirl motion takes place in the other direction (backward whirl motion) with the same rotation of the gyro rotor, we get the corresponding equations by substituting $\nu \to -\nu$. Thus, we get from (142) and (143) in easily comprehensible symbols :

(144)
$$\overset{\curvearrowright}{x} = e_r \cos \nu t$$
$$\overset{\curvearrowright}{y} = -e_r \sin \nu t$$

Self-Supporting Bearings for the Spin-Axis

$$K_{\vec{x}} = (-k_{\vec{\parallel}} \cos vt + k_{\vec{\perp}} \sin vt) e_r$$

$$K_{\vec{y}} = (k_{\vec{\parallel}} \sin vt + k_{\vec{\perp}} \cos vt) e_r \ . \quad\quad (145)$$

We shall now consider a linear harmonic vibration of the gyro-rotor in the x direction ($x = e_r \cos vt$, $y = 0$). This motion can be synthetized as the algebraic average of the forward and backward circular motions. If we write the corresponding components of the forces with K_{xx} and K_{yx} we get:

$$K_{xx} = \tfrac{1}{2}(K_{\vec{x}} + K_{\vec{x}})$$

$$K_{yx} = \tfrac{1}{2}(K_{\vec{y}} + K_{\vec{y}}) \ . \quad\quad (146)$$

If we introduce the relations (145) we get:

$$K_{xx} = -\tfrac{1}{2} e_r \operatorname{Real}\left\{[(k_{\vec{\parallel}} + k_{\vec{\parallel}}) - i(k_{\vec{\perp}} - k_{\vec{\perp}})] e^{ivt}\right\} \quad (147)$$

$$K_{yx} = \tfrac{1}{2} e_r \operatorname{Real}\left\{[(k_{\vec{\perp}} + k_{\vec{\perp}}) + i(k_{\vec{\parallel}} - k_{\vec{\parallel}})] e^{ivt}\right\} \ . \quad (148)$$

To compare the above with eq. (127) we write, taking into account eq. (128) and (129), in a complex representation:

$$\begin{bmatrix} K_{xx} \\ K_{yx} \end{bmatrix} = -e_r \operatorname{Real} \begin{bmatrix} b_{\parallel} & b_{\perp} \\ -b_{\perp} & b_{\parallel} \end{bmatrix} \begin{bmatrix} e^{ivt} \\ 0 \end{bmatrix} \ . \quad\quad (149)$$

Using the equation (136) we get:

$$(150) \qquad \begin{bmatrix} K_{xx} \\ K_{yx} \end{bmatrix} = -e_r \operatorname{Real} \begin{bmatrix} (u_\| + iv_\|)e^{ivt} \\ -(u_\perp + iv_\perp)e^{ivt} \end{bmatrix}.$$

The comparison of eq. (147) and (148) with eq. (150) yields:

$$(151) \qquad \left. \begin{array}{ll} u_\| = \frac{1}{2}(k_{r\|} + k_{T^*}) \; ; & v_\| = -\frac{1}{2}(k_{\perp} - k_{T}) \\ u_\perp = \frac{1}{2}(k_{\perp} + k_{T}) \; ; & v_\perp = \frac{1}{2}(k_{r\|} - k_{T^*}) \end{array} \right\}.$$

From (137) and (138) follow the conditions for the instability threshold:

$$(152) \qquad m_{cr} = \frac{1}{v_{cr}^2} \left\{ \begin{array}{l} k_{r\|} \\ k_{T} \end{array} \right|_{v_{cr}}$$

$$(153) \qquad \left. \begin{array}{l} -k_\perp \\ k_{T^*} \end{array} \right\}\bigg|_{v_{cr}} = 0 \; .$$

The equations (153) yield the critical frequences for the forward and backward whirl motion, the equations (152) yield the corresponding critical masses. We recognize that the critical states are reached if the components of forces perpendicular to the displacement vanish. The components of forces in the direction of the displacement will be balanced according to (152) by the centrifugal forces of the circular whirl motion.

From eq. (141) and (151) we get the values of λ' for the two whirl motions if m' is the deviation from the critical mass. For the forward whirl motion we have:

$$\lambda' = \left\{ \frac{-v^2 \frac{\partial k_{\uparrow\uparrow}}{\partial v} m'}{\left[v^2 \frac{\partial}{\partial v}\left(\frac{k_{\uparrow\uparrow}}{v^2}\right) \right]^2 + \left[\frac{\partial k_{\uparrow\uparrow}}{\partial v} \right]^2} \right\}\Bigg|_{v_{cr}} \quad (154)$$

and for the backward whirl motion we have:

$$\lambda' = \left\{ \frac{v^2 \frac{\partial k_{\uparrow}^*}{\partial v} m'}{\left[v^2 \frac{\partial}{\partial v}\left(\frac{k_{\uparrow}^*}{v^2}\right) \right]^2 + \left[\frac{\partial k_{\uparrow}^*}{\partial v} \right]^2} \right\}\Bigg|_{v_{cr}} \quad (155)$$

From the above we recognize that we have an unstable field ($\lambda' > 0$) in case of a positive mass excess ($m' > 0$), if there is $\frac{\partial k_{\uparrow\uparrow}}{\partial v}\Big|_{v_{cr}} < 0$ in case of the forward whirl motion, and if there is $\frac{\partial k_{\uparrow}^*}{\partial v}\Big|_{v_{cr}} > 0$ in case of the backward whirl motion. With the negative change of mass ($m' < 0$) it is the contrary.

References.

1. Ostwalds Klassiker, Nr. 218. Leipzig, Akademische Verlagsges. 1927.
2. G. Hirn : Sur les principaux phénomènes qui présentent les frottements médiats. Soc. ind. Mulhouse Bull., 26 (1854).
3. W.J. Harrison : The Hydrodynamical Theory of Lubrication with Special Reference to Air as a Lubricant. Trans. Cambridge Phil. Soc. 22 (1913).
4. A. Wiemer, Luftlagerung, Veb Verlag Technik Berlin, 1969.
5. G. Heinrich, Über Strönmungslager. Maschbau. u. Warmew. 4 (1949).
6. G. Heinrich, Das aerodynamische Lager. Maschbau. u. Warmew. 7 (1952).
7. S.F. Murray and M.B. Peterson, The Selection and Evaluation of Materials and Lubricant Films for Gas-Lubricated Gyro Bearings, MTI 64 TRI, prepared for Special Projects Office SP 24, US Navy, Contract Nobs-88615 (FBM), Mechanincal Technology Incorporates Latham, New York.
8. C.H.T. Pan, Spectral Analysis of Gas Bearing Systems

for Stability Studies, Developments in Mechanics, Aug. 16-18, 1955, University of Wisconsin, ed. T. C. Huang and M. W. Johnson, John Wiley and Sons, Inc., New York, N. Y. 1967.

9. S. B. Malanoski, Journal Bearing Dynamic Response, MTI 67 TR 88, Mechanical Technology Incorporates, Lathan, N. Y.

10. J. H. Vohr and C. H. T. Pan, Design Data Gas. Lubricated Spin-Axis Bearings for Gyroscopes, MTI-68 TR 29, Mechanical Technology Incorporates, Lathan, N. Y.

Contents.

	Page
Foreword	3
1. Introduction	5
2. Parallel Flow in the Gap	7
3. Applications to Statical Gas Bearings	22
4. Self-Supporting Bearings for the Spin-Axis	38
References	55

MIX
Papier aus verantwortungsvollen Quellen
Paper from responsible sources
FSC® C105338

If you have any concerns about our products,
you can contact us on
ProductSafety@springernature.com

In case Publisher is established outside the EU,
the EU authorized representative is:
**Springer Nature Customer Service Center GmbH
Europaplatz 3, 69115 Heidelberg, Germany**

Printed by Libri Plureos GmbH
in Hamburg, Germany